Mathematics
Team Races

Seventeen ready-to-use activities to make learning more effective and more fun!
Paul Hambleton

	Introduction	3
Race 1	Percentage Problems	4
Race 2	Solving Quadratics by Factorising	6
Race 3	Solving Quadratic Equations by Formula	8
Race 4	Adding and Subtracting Algebraic Equations	10
Race 5	Multiplying and Dividing Algebraic Fractions	12
Race 6	Direct Proportion	14
Race 7	Inverse Proportion	16
Race 8	Solving Equations with Fractions	18
Race 9	Rearranging Formulae	20
Race 10	Linear Simultaneous Equations	22
Race 11	Non-linear Simultaneous Equations	24
Race 12	nth Term of Quadratic Sequences	26
Race 13	Pythagoras' Theorem	28
Race 14	Right-angled Trigonometry	30
Race 15	The Sine and Cosine Rules	32
Race 16	Solving Trigonometric Equations	34
Race 17	Probabilty	36
	Templates	38

tarquin

Welcome to
Mathematical Team Races

Team Races are special mathematical puzzles and problems which demand and produce real co-operation among the members of the team.

The team races included in this collection are each presented in the form of photocopiable race cards. Pairs works very well, giving students enough to do, but larger teams can work well too. The number of teams you decide on is the number of copies of the photocopiable card page you will need.

Cut out the cards and arrange into stacks of Round 1, Round 2, etc. in front of you. Give each team the card with Round 1 on it. You may prefer to place these face down and get students to turn them over at the same time, which can help to create a buzz about the activity.

All solutions to the puzzles are provided together with suggestions about how the answers might be organised and presented.

Paul Hambleton

Distributed in the USA by Parkwest
www.parkwestpubs.com
www.amazon.com & major retailers

Distributed in Australia by OLM
www.lat-olm.com.au

© 2013 Paul Hambleton
ISBN: 978 1 907550 21 8
Printed and designed in the UK

www.tarquingroup.com

All rights reserved
Photocopiable under Schools
Licences only

tarquin publications
Suite 74, 17 Holywell Hill
St Albans, AL1 1DT, UK

Introduction

A team race is a classic idea (your team wins by answering all the questions correctly in the fastest time!) that I used in a lesson once and it worked so well it became something I did whenever that topic came up, as students enjoyed the activity so much, and I felt the quality of learning and   discussion taking place was very high. However, while I could find races for general mathematics (great for end of term activities), I couldn't find many resources for specific topics so I could use it more and with different classes, and that is what this book contains. Whenever I have used this activity student feedback has always been very positive. The element of competition means students must answer questions as quickly as possible, but accuracy is also essential as students need correct answers in order to progress and win the race.

How to organise a team race

Decide what size teams you would like students to be in. Pairs works very well, giving students enough to do, but larger teams can work well too. The number of teams you decide on is the number of copies of the photocopiable card page you will need. Cut out the cards and arrange into stacks of Round 1, Round 2, etc. in front of you. Give each team the card with Round 1 on it. You may prefer to place these face down and get students to turn them over at the same time, which can help to create a buzz about the activity.

Students then answer the question on Round 1, and when they have an answer they present the answer to you (ensuring they don't allow other pairs or groups to see the answer). You may like to get students to write answers on paper, in books, or on mini-whiteboards. If the answer is correct, give them Round 2, and this continues until the first pair or group has completed all the rounds you wish them to complete. It may be necessary to offer some quiet support to some pairs or groups if they are struggling on a particular round, otherwise they may become stuck and despondent. The winning pair or group is the first to complete all rounds.

You don't have to use all 10 rounds. The questions are written with the intention that in general, each round is more difficult than the last. So you can have a quick glance at the questions and decide which rounds you want to use in your team race. You may miss off the first few if they are too easy, or the last few if they are too difficult.

You may wish to print the cards on coloured card or laminate them to make them more durable. Once copied, as long as you don't allow students to write on the cards, they can be used again and again, and the activity is very simple to set up, as the cards and something to write with are the only resources you need.

Answers

An answer sheet is provided for each of the pre-prepared activities. Make sure students can't see these!

Active learning

Active learning strategies are becoming ever more popular in schools wishing to give students not only a more enjoyable experience but also a deeper understanding of mathematics through collaborative learning and more independent learning, and this type of activity can encourage this.

Differentiation

These activities are easy to differentiate. You can miss off later, more difficult rounds as appropriate to the ability of your students, or you can easily make up some more challenging questions to tack on as additional rounds at the end. In particular often the last few rounds include worded questions, where students have to interpret the question and use the skills developed in an earlier part of the race to answer them. As this is a collaborative activity, students will be able to support each other.

Make your own versions

Once you have used some of the pre-prepared sets of cards, you may wish to create your own versions. There is a blank template provided for this purpose.

Which topics does this work well for?

Team races work well for topics which require some degree of working out to be done by students, for two reasons. Firstly for the activity to work well, the teacher needs to check each answer, and so there should be enough work on each question to keep students busy for a few minutes. Secondly, this idea promotes collaborative working and multi-stage questions are ideally suited to this.

Race 1

Percentage Problems

Topics
percentage increase and decrease
finding percentage change
simple and compound interest
depreciation
reverse percentages

TEACHER'S NOTES

This activity is a great consolidation exercise in percentage problems, including reverse percentages, compound interest and depreciation. There are several percentages skills here, but more importantly, students will need to determine which method is appropriate for each question.

ANSWERS

Round 1	Round 2
4%	25%, £270
Round 3	**Round 4**
20%	10%
Round 5	**Round 6**
£528	£4,433.70
Round 7	**Round 8**
£8,153.35	£95
Round 9	**Round 10**
£180,000	£8,999.99

 **Round 1**

 **Round 2**

Mary's salary increases from £22,500 per year to £23,400 per year. By what percentage has her salary increased?

Mark buys 18 bicycles for £60 and sells them each for £75. What is the percentage profit that he has made on each bicycle, and how much total profit has he made?

 **Round 3**

 **Round 4**

A used car salesman buys a car for £4,000 but he has overpaid for the car and he can only sell it for a maximum price of £3,200. What percentage loss has he made?

A chocolate manufacturer reduces the size of its standard bar during a re-design, from 20 squares of chocolate to 18 squares. What percentage of chocolate has been lost from each bar?

 **Round 5**

 **Round 6**

Marissa invests her £400 savings in an account offering simple interest of 8% p.a. for 4 years. How much does she have in her account at the end of the 4 years?

Mario pays a £3,500 bonus into a compound interest account paying 3% p.a. He makes no withdrawals; how much will be in his account after 8 years?

 **Round 7**

 **Round 8**

Martha buys a new car for £12,995. She estimates that it will depreciate in value at a rate of 11% per year. How much does she estimate it will be worth after 4 years?

A furniture salesman increases his prices by 4% as he thinks this is roughly the rate of inflation. The new price of a chair is £98.80. How much did it cost before the increase?

 **Round 9**

 **Round 10**

A homeowner has dropped the asking price of their house by 15%. The new asking price is £153,000. What was the original asking price?

Matthew's 5-year old car is now valued at £5931.73. He calculates it has depreciated by 8% each year since he bought it. How much was it when he bought it?

Race 2

Solving Quadratic Equations by Factorising

Topics
solving quadratic equations by factorising (coefficient of x² is 1)
solving quadratic equations by factorising (coefficient of x² is greater than 1)

TEACHER'S NOTES

You can decide whether you want to include the questions where the coefficient of x² is greater than 1 or not, depending on the ability/experience of your students.

ANSWERS

Round 1	Round 2
$x = -1$ or $x = -2$	$x = -4$ or $x = -9$
Round 3	**Round 4**
$x = -1$ or $x = 8$	$x = -5$ or $x = 7$
Round 5	**Round 6**
$x = 4$ (repeated root)	$x = 1$ or $x = 3$
Round 7	**Round 8**
$x = 2$ or $x = 6$	$x = \frac{1}{2}$ or $x = 2$
Round 9	**Round 10**
$x = \frac{3}{2}$ or $x = 5$	$x = -1$ or $x = \frac{5}{2}$

Answers could appear as decimal or fraction form for questions where the coefficient of x² is greater than 1. You will either need to prime students beforehand to give answers in a particular preferred format, or be willing to accept either as the solution.

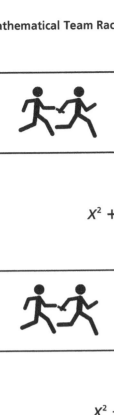 **Round 1**

Solve

$x^2 + 3x + 2 = 0$

 Round 2

Solve

$x^2 + 13x + 36 = 0$

 **Round 3**

Solve

$x^2 - 7x - 8 = 0$

 Round 4

Solve

$x^2 - 2x - 35 = 0$

 Round 5

Solve

$x^2 - 8x + 16 = 0$

 Round 6

Solve

$x^2 - 4x + 3 = 0$

 Round 7

Solve

$x^2 - 8x + 12 = 0$

 Round 8

Solve

$2x^2 - 5x + 2 = 0$

 Round 9

Solve

$2x^2 - 13x + 15 = 0$

Round 10

Solve

$2x^2 - 3x - 5 = 0$

Race 3

Solving Quadratic Equations by Formula

Topics
solving quadratic equations using the formula
solving word problems by forming and solving a quadratic equation

TEACHER'S NOTES

Students will need to have met the Quadratic Formula and know how to use it. Other than that, this is a great way to learn or consolidate this topic as students will have to ensure they are substituting into the formula correctly, and if they are incorrect, it will usually be possible for students to re-attempt until they have the correct solution. You could also use this activity to enable students to practise solving quadratic equations by completing the square.

ANSWERS

Round 1	Round 2
$x = 1.24$ or $x = -3.24$	$x = 0.69$ or $x = -2.19$
Round 3	Round 4
$x = 0.41$ or $x = -2.41$	$x = 0.72$ or $x = -1.12$
Round 5	Round 6
$x = 1.43$ or $x = 0.23$	$x = 3.16$ or $x = -0.16$
Round 7	Round 8
$x = 1.10$ and $x = -2.43$	No real roots
Round 9	Round 10
The width (x) is 5.59m	$x = 5.29$ so the width is $x + 29 =$ 5.29 + 29 = 34.29

Answers here are given to 2 decimal places. You may prefer students to give answers in surd form, but whatever you decide, they will need to know what answer format you will accept. It could be a good exercise in highlighting the importance of giving the answer in the form asked for in the question, or you may decide to relax this issue and allow students to focus on using the formula correctly.

 **Round 1**

Solve
$$x^2 + 2x - 4 = 0$$

 **Round 2**

Solve
$$2x^2 + 3x - 3 = 0$$

 **Round 3**

Solve
$$x^2 + 2x - 1 = 0$$

 **Round 4**

Solve
$$5x^2 + 2x - 4 = 0$$

 **Round 5**

Solve
$$3x^2 - 5x + 1 = 0$$

 **Round 6**

Solve
$$2x^2 - 6x - 1 = 0$$

 **Round 7**

Solve
$$3x^2 + 4x - 8 = 0$$

 **Round 8**

Solve
$$4x^2 - 3x + 4 = 0$$

 **Round 9**

The length of a rectangular field is 3 metres more than its width. Its area is 48 m^3. Form and solve a quadratic equation in x to find the width.

 **Round 10**

The height of a particular TV screen is $(2x + 15)$ inches. The width is $(x + 29)$ inches. The area of the TV is 877 square inches. Find the width of the TV.

Race 4

Adding and Subtracting Algebraic Fractions

Topics
adding algebraic fractions
subtracting algebraic fractions

TEACHER'S NOTES

Ensure students are confident dealing with simple examples first, and have an accurate method to follow with harder examples, as in this activity. As long as they are confident with a valid method, the only difficulties which may arise are the expanding and simplifying involved in the numerators, which students should be able to work out in their pairs/teams.

ANSWERS

Round 1	Round 2
$$\dfrac{3(x + 1)}{x(x + 3)}$$	$$\dfrac{7x + 2}{(x + 1)(2x - 3)}$$
Round 3	Round 4
$$\dfrac{2x^2 + 4x + 4}{(x + 4)(2x + 3)}$$	$$\dfrac{-x^2 + 2x + 5}{x^2 - 1}$$
Round 5	Round 6
$$\dfrac{6x^2 + 15x - 5}{3x(3x - 5)}$$	$$\dfrac{x^2 + 5x + 30}{(x - 9)(x + 3)}$$
Round 7	Round 8
$$\dfrac{11x + 6}{x^2 - 4}$$	$$\dfrac{1}{x + 3}$$
Round 9	Round 10
$$\dfrac{x^2 + x + 2}{x^2(x + 2)}$$	$$\dfrac{2x^3 + x - 8}{x^2(x - 8)}$$

The expressions in the numerators and denominators of the answers can be expressed in several forms, and you will either need to guide students as to which format to give (always factorised fully, always expanded fully) or be prepared to accept either. However, students should in all cases have simplified like terms as far as possible.

	Round 1		Round 2
$\dfrac{1}{x} + \dfrac{2}{x+3}$		$\dfrac{1}{x+1} + \dfrac{5}{2x-3}$	
	Round 3		Round 4
$\dfrac{1}{2x+3} + \dfrac{x}{x+4}$		$\dfrac{3}{x-1} - \dfrac{x+2}{x+1}$	
	Round 5		Round 6
$\dfrac{1}{3x} + \dfrac{2x+4}{3x-5}$		$\dfrac{x+4}{x-9} - \dfrac{2}{x+3}$	
	Round 7		Round 8
$\dfrac{4}{x+2} + \dfrac{7}{x-2}$		$\dfrac{6}{x+3} - \dfrac{5}{x+3}$	
	Round 9		Round 10
$\dfrac{1}{x^2} + \dfrac{1}{x+2}$		$\dfrac{1}{x^2} + \dfrac{2x}{x-8}$	

Race 5

Multiplying and Dividing Algebraic Fractions

Topics
index laws
multiplying algebraic fractions
dividing algebraic fractions
simplifying algebraic fractions

TEACHER'S NOTES

Students will need to be familiar with the laws of indices and they will need to know how to multiply numerical fractions. The last two rounds have dividing questions.

ANSWERS

Round 1	Round 2
$\dfrac{1}{xy}$	c
Round 3	**Round 4**
$\dfrac{xy^2z}{4}$	$\dfrac{15x^7}{4y^6z}$
Round 5	**Round 6**
$\dfrac{8b^2}{3a^3c^5}$	$\dfrac{27p}{r^6}$
Round 7	**Round 8**
$\dfrac{6b^4c}{5a}$	$\dfrac{y^5}{x^5} = \left(\dfrac{y}{x}\right)^5$
Round 9	**Round 10**
$\dfrac{y^3}{z^3} = \left(\dfrac{y}{z}\right)^3$	$\dfrac{2}{3a}$

Depending on your students, you may wish to be quite strict in only allowing answers where variables are written in alphabetical order.

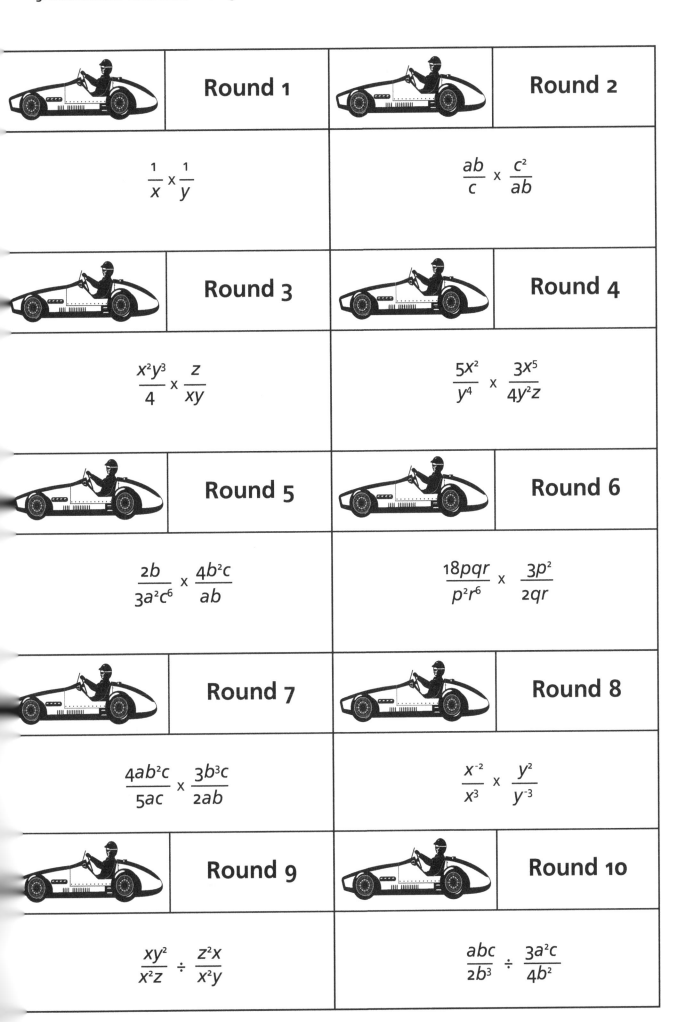

Round 1

$$\frac{1}{x} \times \frac{1}{y}$$

Round 2

$$\frac{ab}{c} \times \frac{c^2}{ab}$$

Round 3

$$\frac{x^2y^3}{4} \times \frac{z}{xy}$$

Round 4

$$\frac{5x^2}{y^4} \times \frac{3x^5}{4y^2z}$$

Round 5

$$\frac{2b}{3a^2c^6} \times \frac{4b^2c}{ab}$$

Round 6

$$\frac{18pqr}{p^2r^6} \times \frac{3p^2}{2qr}$$

Round 7

$$\frac{4ab^2c}{5ac} \times \frac{3b^3c}{2ab}$$

Round 8

$$\frac{x^{-2}}{x^3} \times \frac{y^2}{y^{-3}}$$

Round 9

$$\frac{xy^2}{x^2z} \div \frac{z^2x}{x^2y}$$

Round 10

$$\frac{abc}{2b^3} \div \frac{3a^2c}{4b^2}$$

Race 6

Direct Proportion

Topics
forming equations relating variables in direct proportion
using a proportional relationship formula to calculate one variable when given the other

TEACHER'S NOTES

In each of these rounds, students must derive a formula linking the two variables in question. They must also then calculate the value of one variable, given the other, using the formula that they have derived. This activity is all about getting students to translate the proportional relationship in words into a symbolic formula, and then use that formula by substituting in the given values.

ANSWERS

Round 1	Round 2

$y = 2x$
    $y = 40$ when $x = 20$
	$y = 6x$
	    $x = 8$ when $y = 48$

Round 3	**Round 4**

$y = 3x^2$
    $y = 75$ when $x = 5$
	$y = 2x^2$
	    $x = 7$ when $y = 98$

Round 5	**Round 6**

$q = 4\sqrt{p}$
    $q = 16$ when $p = 16$
	$q = 5\sqrt{p}$
	    $p = 9$ when $q = 15$

Round 7	**Round 8**

$y = 4x^3$
    $y = 108$ when $x = 3$
	$b = 2a^3$
	    $a = 4$ when $b = 128$

Round 9	**Round 10**

$y = 5\sqrt[3]{x}$
    $y = 15$ when $x = 27$
	$y = 3\sqrt[3]{x}$
	    $x = 8$ when $y = 6$

	Round 1		Round 2
	y is proportional to *x* and when *x* = 4, *y* = 8. Find a formula for *y* in terms of *x*. What is the value of *y* when *x* = 20?		*y* is proportional to *x* and when *x* = 1, *y* = 6. Find a formula for *y* in terms of *x*. What is the value of *x* when *y* = 48?
	Round 3		Round 4
	y is proportional to the square of *x* and when *x* = 2, *y* = 12. Find a formula for *y* in terms of *x*. What is the value of *y* when *x* = 5?		*y* is proportional to the square of *x* and when *x* = 4, *y* = 32. Find a formula for *y* in terms of *x*. What is the value of *x* when *y* = 98?
	Round 5		Round 6
	q is proportional to the square root of *p* and when *p* = 4, *q* = 8. Find a formula for *q* in terms of *p*. What is the value of *q* when *p* = 16?		*q* is proportional to the square root of *p* and when *p* = 16, *q* = 20. Find a formula for *q* in terms of *p*. What is the value of *p* when *q* = 15?
	Round 7		Round 8
	y is proportional to the cube of *x* and when *x* = 2, *y* = 32. Find a formula for *y* in terms of *x*. What is the value of *y* when *x* = 3?		*b* is proportional to the cube of *a* and when *a* = 3, *b* = 54. Find a formula for *b* in terms of *a*. What is the value of *a* when *b* = 128?
	Round 9		Round 10
	y is proportional to the cube root of *x* and when *x* = 8, *y* = 10. Find a formula for *y* in terms of *x*. What is the value of *y* when *x* = 27?		*y* is proportional to the cube root of *x* and when *x* = 64, *y* = 12. Find a formula for *y* in terms of *x*. What is the value of *x* when *y* = 6?

Race 7

Inverse Proportion

Topics
forming equations relating variables in inverse proportion
using an inversely proportional relationship formula to calculate one variable when given the other

TEACHER'S NOTES

This activity starts with some worded, real-life problems which students should be able to solve quite easily with a bit of logical thinking. The activity then progresses on to more formal situations where a formula must be found, as in the activity on Direct Proportion.

ANSWERS

Round 1	Round 2
30 days	3 hours

Round 3	Round 4
$y = \dfrac{4}{x}$ When $x = 1$, $y = 4$	$y = \dfrac{3}{x}$ When $y = 3$, $x = 1$

Round 5	Round 6
$y = \dfrac{2}{x^2}$ When $x = 2$, $y = 0.5$	$y = \dfrac{5}{x^2}$ When $y = 5$, $x = 1$

Round 7	Round 8
$y = \dfrac{4}{\sqrt{x}}$ When $x = 64$, $y = 0.5$	$y = \dfrac{2}{\sqrt{x}}$ When $y = 4$, $x = 0.25$

Round 9	Round 10
$y = \dfrac{2}{x^3}$ When $x = 3$, $y = 0.074$ (to 3dp)	$y = \dfrac{16}{x^3}$ When $y = 32$, $x = 0.794$ (to 3dp)

 **Round 1**

It takes 3 builders 20 days to build a wall. How long will it take 2 builders?

 **Round 2**

It takes 4 people 9 hours to paint a house. How long will it take 12 people to paint the same house?

 **Round 3**

y is inversely proportional to *x* and when *x* = 2, *y* = 2. Find a formula for *y* in terms of *x*. What is the value of *y* when *x* = 1?

 **Round 4**

y is inversely proportional to *x* and when *x* = 3, *y* = 1. Find a formula for *y* in terms of *x*. What is the value of *x* when *y* = 3?

 **Round 5**

y is inversely proportional to the square of *x* and when *x* = 4, *y* = 0.125. Find a formula for *y* in terms of *x*. What is the value of *y* when *x* = 2?

 **Round 6**

y is inversely proportional to the square of *x* and when *x* = 5, *y* = 0.2. Find a formula for *y* in terms of *x*. What is the value of *x* when *y* = 5?

 **Round 7**

y is inversely proportional to the square root of *x* and when *x* = 16, *y* = 1. Find a formula for *y* in terms of *x*. What is the value of *y* when *x* = 64?

 **Round 8**

y is inversely proportional to the square root of *x* and when *x* = 4, *y* = 1. Find a formula for *y* in terms of *x*. What is the value of *x* when *y* = 4?

 **Round 9**

y is inversely proportional to the cube of *x* and when *x* = 2, *y* = 0.25. Find a formula for *y* in terms of *x*. What is the value of *y* when *x* = 3?

 **Round 10**

y is inversely proportional to the cube of *x* and when *x* = 2, *y* = 2. Find a formula for *y* in terms of *x*. What is the value of *x* when *y* = 32?

Race 8

Solving Equations with Fractions

Topics
manipulating algebraic equations including fractions
solving quadratic equations

TEACHER'S NOTES

Students will need to be able to manipulate equations and expressions fluently, and they will also need to be confident solving quadratic equations using the formula or by completing the square.

ANSWERS

Round 1	Round 2
$x = 3$	$x = 5$
Round 3	Round 4
$x = 7.45$ or -6.45	$x = 5$
Round 5	Round 6
$x = 4.55$ or -0.55	$x = 1$ or $\frac{5}{4}$
Round 7	Round 8
$x = 0.58$ or -1.63	$x = 1.86$ or -1.22
Round 9	Round 10
$x = -1.48$ or 3.15	$x = 3$ or -0.40

Where appropriate, answers have been given to 2 decimal places.

Round 1

Solve
$$\frac{2x + 1}{x - 2} = 7$$

 Round 2

Solve
$$\frac{3x - 3}{x + 1} = 2$$

 Round 3

Solve
$$\frac{6}{x + 2} = \frac{x - 3}{7}$$

 Round 4

Solve
$$\frac{x + 1}{x - 3} = \frac{x + 4}{x - 2}$$

 Round 5

Solve
$$\frac{1}{2x + 1} + \frac{5}{x + 1} = 1$$

 Round 6

Solve
$$\frac{2}{x - 3} - \frac{3}{2x + 1} = -2$$

 Round 7

Solve
$$\frac{x}{2x - 1} + \frac{2x}{3x + 4} = 4$$

 Round 8

Solve
$$\frac{x}{x + 1} - \frac{3x}{2x - 5} = 5$$

 Round 9

Solve
$$\frac{4x}{x + 2} + \frac{x + 2}{2x + 3} = 3$$

 Round 10

Solve
$$\frac{1}{x + 1} + \frac{1}{x - 2} = \frac{5}{4}$$

Race 9

Rearranging Formulae

Topics
rearranging formulae where the new subject appears once

TEACHER'S NOTES

In all of the formulae in this activity, the variable which is to become the new subject appears once. They are mostly area or volume formulae, and some will be familiar to students. A good extension activity could be to give bonus points if they can also tell you the shape whose area or volume is given by the formula.

ANSWERS

Round 1	Round 2
$L = \dfrac{A}{W}$	$d = \dfrac{A}{\pi}$
Round 3	**Round 4**
$r = \dfrac{A}{\pi l}$	$t = \dfrac{v - u}{a}$
Round 5	**Round 6**
$h = \dfrac{3V}{\pi r^2}$	$r = \sqrt{\dfrac{A}{\pi}}$
Round 7	**Round 8**
$h = \dfrac{2A}{b}$	$h = \dfrac{2A}{a + b}$
Round 9	**Round 10**
$r = \sqrt{\dfrac{A}{4\pi}}$	$r = \sqrt[3]{\dfrac{3V}{4\pi}}$

As all of the new subjects in these formula refer to dimensions of areas or volumes, negative square roots have not been included in the answers, but you may feel this is an area for discussion with students either before or after the activity.

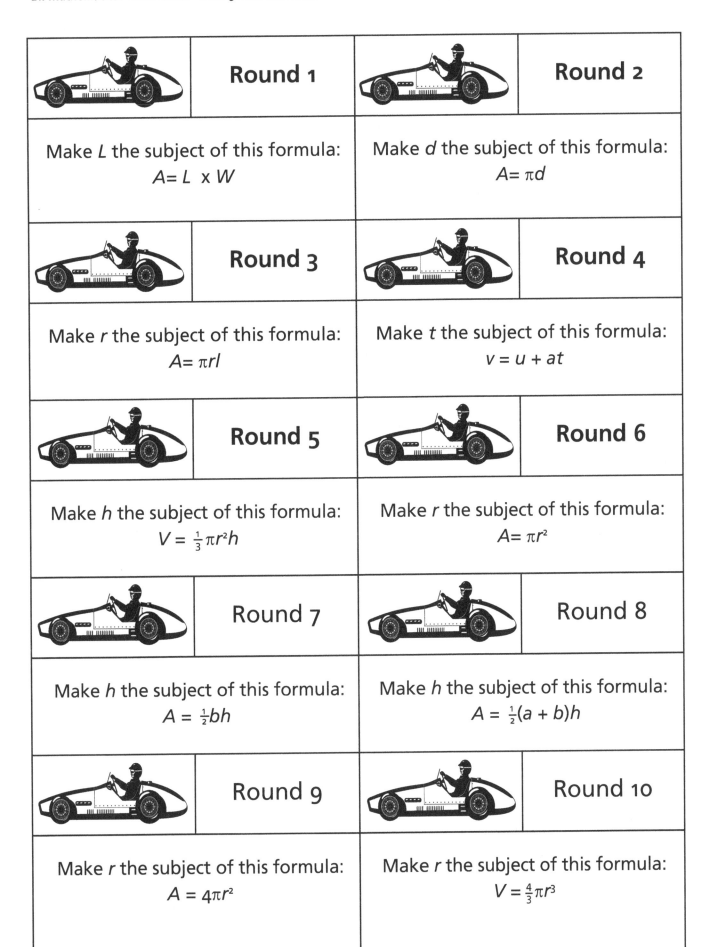

Round 1

Make *L* the subject of this formula:
$$A = L \times W$$

Round 2

Make *d* the subject of this formula:
$$A = \pi d$$

Round 3

Make *r* the subject of this formula:
$$A = \pi r l$$

Round 4

Make *t* the subject of this formula:
$$v = u + at$$

Round 5

Make *h* the subject of this formula:
$$V = \tfrac{1}{3}\pi r^2 h$$

Round 6

Make *r* the subject of this formula:
$$A = \pi r^2$$

Round 7

Make *h* the subject of this formula:
$$A = \tfrac{1}{2}bh$$

Round 8

Make *h* the subject of this formula:
$$A = \tfrac{1}{2}(a + b)h$$

Round 9

Make *r* the subject of this formula:
$$A = 4\pi r^2$$

Round 10

Make *r* the subject of this formula:
$$V = \tfrac{4}{3}\pi r^3$$

Race 10

Linear Simultaneous Equations

Topics
solving linear simultaneous equations by elimination
forming and solving linear simultaneous equations

TEACHER'S NOTES

All of these systems of linear simultaneous equations can be solved using the method of elimination. The substitution method is covered in the next race. There are questions covering the full range of different types; where neither equation needs to be multiplied, where one does and where both do. There are also some questions where adding will eliminate as well as some where subtracting will eliminate.

ANSWERS

Round 1	Round 2
(1,2)	(2,3)
Round 3	**Round 4**
(1,4)	(-2,1)
Round 5	**Round 6**
(1,-5)	(0.5,2)
Round 7	**Round 8**
(1.5,-2)	£25
Round 9	**Round 10**
150 single tickets and 250 return tickets	2 CDs and 5 DVDs

Let students know beforehand what format you want the answer to take. I have given the solutions as co-ordinates of the intersection of the corresponding graphs, but if you have just been working on solving equations rather than graphs, you will probably like to accept $x =...$, $y =...$ as an acceptable format for the solutions.

 **Round 1**

Solve these equations simultaneously
$$x + 2y = 5$$
$$x + 5y = 11$$

 **Round 2**

Solve these equations simultaneously:
$$4x + 3y = 17$$
$$2x + y = 7$$

 Round 3

Solve these equations simultaneously:
$$x + 6y = 25$$
$$5x - 3y = -7$$

 Round 4

Solve these equations simultaneously:
$$2x + 2y = -2$$
$$6x + 7y = -5$$

 Round 5

Solve these equations simultaneously:
$$2x + 3y = -13$$
$$8x - 5y = 33$$

 Round 6

Solve these equations simultaneously:
$$2x - 2y = -3$$
$$4x + 5y = 12$$

 Round 7

Solve these equations simultaneously:
$$6x + y = 7$$
$$2x + 2y = -1$$

 Round 8

3 pencils and 4 pens cost £18.
4 pencils and 3 pens cost £17.

What is the cost of 5 pens and 5 pencils?

 Round 9

A train ticket machine sells single (costing £5) and return tickets (costing £8). One day it sells 400 tickets, taking £2,750.

How many of each type of ticket were sold?

 Round 10

Mark buys 7 items in an outlet shop that only sells CDs and DVDs. CDs cost £6, and DVDs cost £3. Mark's bill comes to £27.

How many CDs and how many DVDs did Mark buy?

Race 11

Non-linear Simultaneous Equations

Topics
solving simultaneous equations where one is linear and one is quadratic
solving simultaneous equations where one is linear and one is the equation of a circle

TEACHER'S NOTES

Your students will need to be confident solving linear simultaneous equations and solving quadratic equations to complete this activity. It is worthwhile linking the solutions of the equations to the graphs, and in doing so students are more likely to remember that there are usually more solutions than when solving linear simultaneous equations. An extension activity following this activity could be to get students to sketch the curves and lines in one or two of the questions, making sure the intersections fit the solutions they found.

ANSWERS

Round 1	Round 2
(2,1)	(1,3)
Round 3	Round 4
(1.56,2.56) (-2.56,-1.56)	(0.76,-1.47) (5.24,7.47)
Round 5	Round 6
(0.47,1.94) (-1.27,-1.54)	(-0.65,-3.95) (1.85,3.55)
Round 7	Round 8
(2.58,-1.16) (-0.58,5.16)	(-3.38,3.69) (4.98,-0.49)
Round 9	Round 10
(1.65,3.65) (-3.65,-1.65)	(2,2)

Answers are given to 2 decimal places where appropriate. Ensure students are aware of the degree of accuracy you expect them to answer to. I have given the solutions as co-ordinates of the intersection of the corresponding graphs, but you may prefer to accept $x =...$, $y =...$ as an acceptable format for the solutions.

Round 1

Solve these equations simultaneously
$$y = 2x - 3$$
$$x + y = 3$$

Round 2

Solve these equations simultaneously:
$$y = 5x - 2$$
$$2x + y = 5$$

Round 3

Solve these equations simultaneously:
$$y = x^2 + 2x - 3$$
$$y = x + 1$$

Round 4

Solve these equations simultaneously:
$$y = x^2 - 4x + 1$$
$$y = 2x - 3$$

Round 5

Solve these equations simultaneously:
$$x^2 + y^2 = 4$$
$$y = 2x + 1$$

Round 6

Solve these equations simultaneously:
$$x^2 + y^2 = 16$$
$$y = 3x - 2$$

Round 7

Solve these equations simultaneously:
$$y = 2x^2 - 6x + 1$$
$$2x + y = 4$$

Round 8

Solve these equations simultaneously:
$$x^2 + y^2 = 25$$
$$x + 2y = 4$$

Round 9

Solve these equations simultaneously:
$$xy = 6$$
$$y = x + 2$$

Round 10

Solve these equations simultaneously:
$$xy = 4$$
$$x + y = 4$$

Race 12

nth Term of Quadratic Sequences

Topics
finding the nth term of a quadratic sequence
calculating later terms of a quadratic sequence

TEACHER'S NOTES

The questions in this race start easy and some students will be able to spot the nth term without too much working. Students will need a method for the later ones, and this exercise is all about getting students to practise the method to become confident finding the nth term. Getting students to check that their expression works for all of the terms in the given sequence is a good idea. If you wish to just keep nth terms where the coefficient of n^2 is 1, just use the first six cards.

ANSWERS

Round 1	Round 2
$n^2 + 2$	$n^2 - 1$
Round 3	Round 4
$n^2 + n$	2650 (nth term is $n^2 + 3n$)
Round 5	Round 6
9800 (nth term is $n^2 - 2n$)	$n^2 + n + 4$
Round 7	Round 8
$2n^2 + n - 1$	$2n^2 + 1$
Round 9	Round 10
5, 14, 27, 44, 65, 90	$3n^2 + 2n + 4$

	Round 1		**Round 2**

Find the *n*th term of this sequence:
3, 6, 11, 18, 27, 38, 51, ...

Find the *n*th term of this sequence:
0, 3, 8, 15, 24, 35, 48, ...

	Round 3		**Round 4**

Find the *n*th term of this sequence:
2, 6, 12, 20, 30, 42, 56, ...

Find the 50th term of this sequence:
4, 10, 18, 28, 40, 54, 70, ...

	Round 5		**Round 6**

What is the 100th term in this sequence?
-1, 0, 3, 8, 15, 24, 35, ...

Find the *n*th term of this sequence:
6, 10, 16, 24, 34, 46, 60, ...

	Round 7		**Round 8**

Find the *n*th term of this sequence:
2, 9, 20, 35, 54, 77, ...

Find the *n*th term of this sequence:
3, 9, 19, 33, 51, 73, 99, ...

	Round 9		**Round 10**

Find the first six terms in the sequence
with *n*th term $2n^2 + 3n$

Find the *n*th term of this sequence:
9, 20, 37, 60, 89, 124, 165, ...

Race 13

Pythagoras' Theorem

Topics
using Pythagoras' Theorem to find the length of the hypotenuse
using Pythagoras' Theorem to find the length of a shorter side
solving word problems using Pythagoras' Theorem

TEACHER'S NOTES

This activity will enable students to practice the full range of skills relating to Pythagoras' Theorem: finding the hypotenuse, finding a shorter side, determining whether a triangle is right-angled or not, and worded problems to solve.

ANSWERS

Round 1	Round 2
8.94 m	6.62 cm
Round 3	Round 4
5.06 cm	11.49 m
Round 5	Round 6
15.09 m	12.81 km
Round 7	Round 8
No You may decide you require working to justify their decision, to avoid guessing.	Yes You may decide you require working to justify their decision, to avoid guessing.
Round 9	Round 10
1.73 m	42 inches

Where appropriate, answers are given to 2 decimal places.

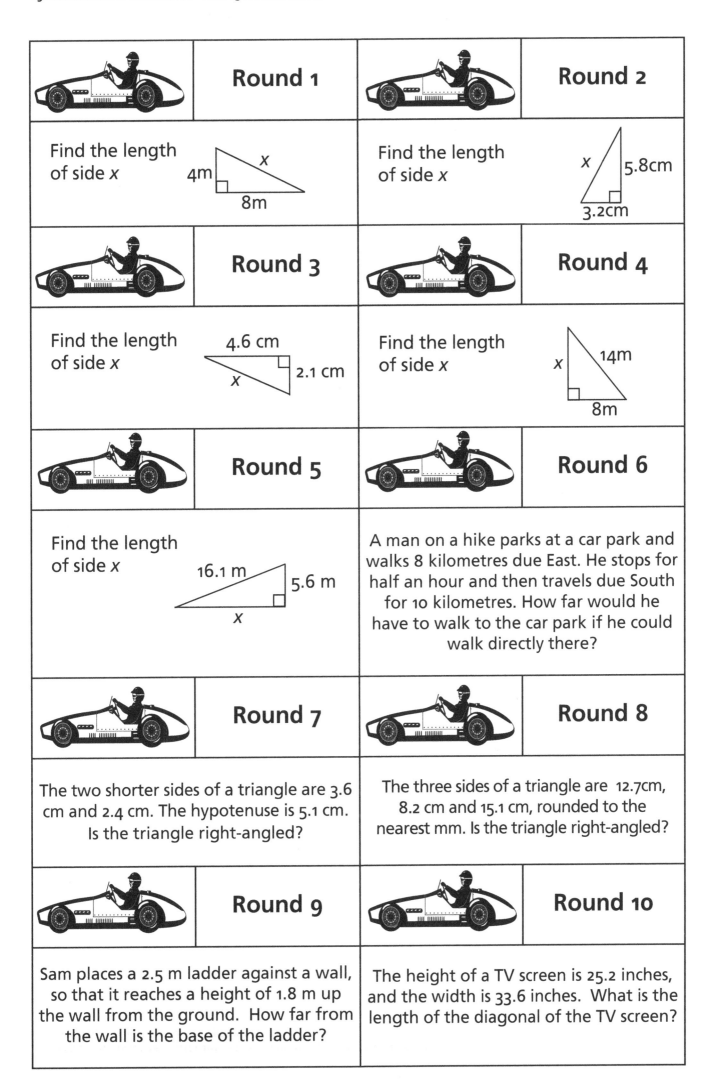

Round 1

Find the length of side *x*

4m, 8m, *x*

Round 2

Find the length of side *x*

x, 5.8cm, 3.2cm

Round 3

Find the length of side *x*

4.6 cm, *x*, 2.1 cm

Round 4

Find the length of side *x*

x, 14m, 8m

Round 5

Find the length of side *x*

16.1 m, 5.6 m, *x*

Round 6

A man on a hike parks at a car park and walks 8 kilometres due East. He stops for half an hour and then travels due South for 10 kilometres. How far would he have to walk to the car park if he could walk directly there?

Round 7

The two shorter sides of a triangle are 3.6 cm and 2.4 cm. The hypotenuse is 5.1 cm. Is the triangle right-angled?

Round 8

The three sides of a triangle are 12.7cm, 8.2 cm and 15.1 cm, rounded to the nearest mm. Is the triangle right-angled?

Round 9

Sam places a 2.5 m ladder against a wall, so that it reaches a height of 1.8 m up the wall from the ground. How far from the wall is the base of the ladder?

Round 10

The height of a TV screen is 25.2 inches, and the width is 33.6 inches. What is the length of the diagonal of the TV screen?

Race 14

Right-angled Trigonometry

Topics
using right-angled trigonometry to find the length of a side
using right-angled trigonometry to find the size of an angle
solving word problems using right-angled trigonometry

TEACHER'S NOTES

This team race contains questions covering all varieties of right-angled trigonometry question. It includes use of the sine, cosine and tangent functions, and questions where students find shorter sides, hypotenuses and angles. There are also two worded questions in the last rounds. It might be worth pointing out to students that none of the diagrams are drawn to scale.

ANSWERS

Round 1	Round 2
$\theta = 36.9°$	$\theta = 39.1°$
Round 3	Round 4
$\theta = 73.4°$	$x = 3.7$ cm
Round 5	Round 6
$x = 5.8$ cm	$x = 7.4$ cm
Round 7	Round 8
$x = 17.4$ m	$x = 8.8$ cm
Round 9	Round 10
The foot of the ladder should be placed 0.7 metres from the wall.	31.0°

All answers are given correct to 1 decimal place.

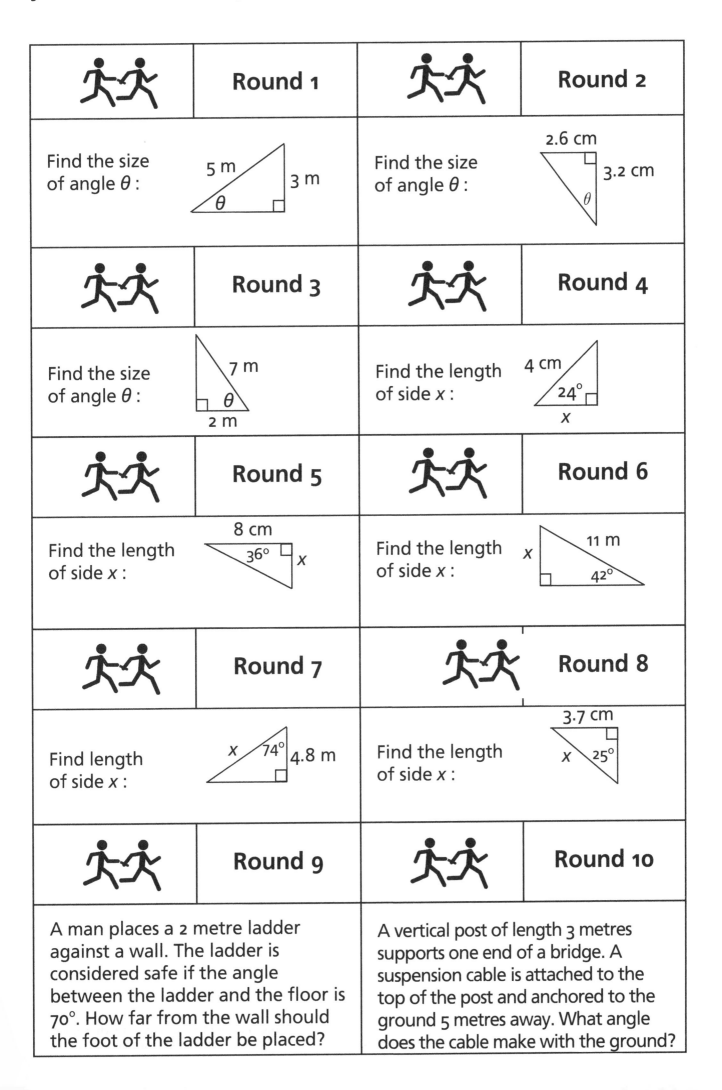

Round 1

Find the size of angle θ :

5 m / 3 m / θ

Round 2

Find the size of angle θ :

2.6 cm / 3.2 cm / θ

Round 3

Find the size of angle θ :

7 m / θ / 2 m

Round 4

Find the length of side x :

4 cm / 24° / x

Round 5

Find the length of side x :

8 cm / 36° / x

Round 6

Find the length of side x :

x / 11 m / 42°

Round 7

Find length of side x :

x / 74° / 4.8 m

Round 8

Find the length of side x :

3.7 cm / x / 25°

Round 9

A man places a 2 metre ladder against a wall. The ladder is considered safe if the angle between the ladder and the floor is 70°. How far from the wall should the foot of the ladder be placed?

Round 10

A vertical post of length 3 metres supports one end of a bridge. A suspension cable is attached to the top of the post and anchored to the ground 5 metres away. What angle does the cable make with the ground?

Race 15

The Sine and Cosine Rules

Topics
using the sine and cosine rules to find the length of a side
using the sine and cosine rules to find the size of an angle
solving word problems using the sine or cosine rule

TEACHER'S NOTES

This activity will allow students to practise using the sine and cosine rules to find missing angles and lengths. It is great for getting students to discuss strategies for solving triangles, and determining from the information given which rule to use. A discussion could follow the activity as to how to identify which questions require which rule. It might be worth pointing out to students that none of the diagrams are drawn to scale.

ANSWERS

Round 1	Round 2
$x = 14.8$ m	$x = 6.8$ m
Round 3	Round 4
$\theta = 91.0°$	$\theta = 52.7°$
Round 5	Round 6
$\theta = 91.8°$	$x = 22.6$ m
Round 7	Round 8
$x = 12.7$ cm	$\theta = 19.5°$
Round 9	Round 10
$\theta = 45.5°$	$x = 6.8$ m

All answers are given correct to 1 decimal place.

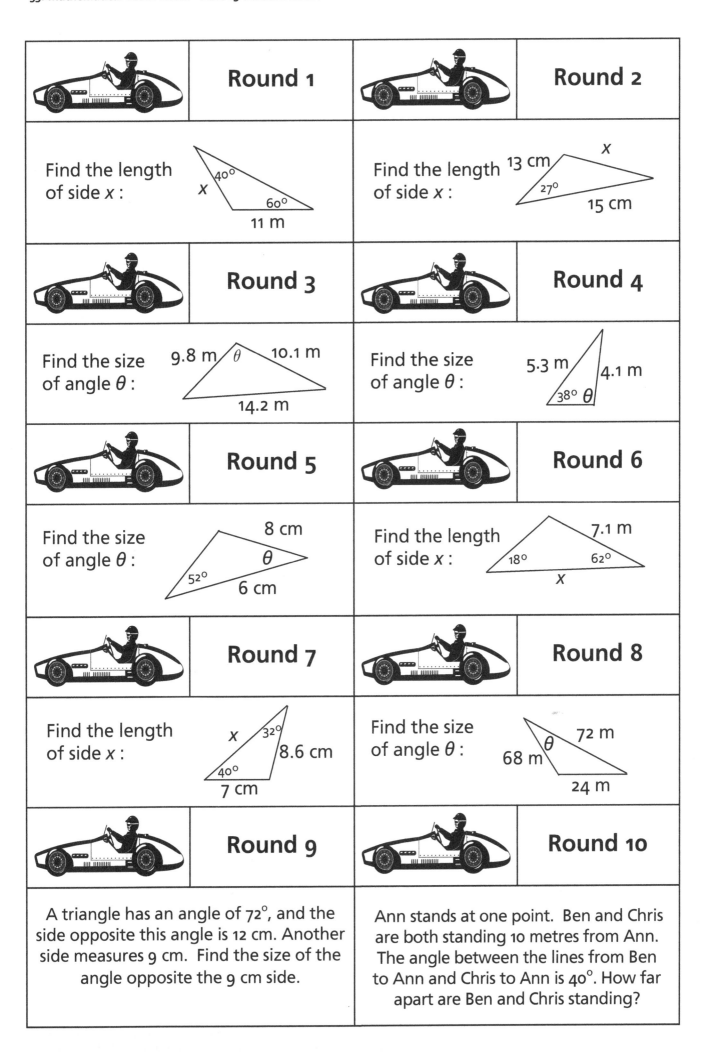

Round 1

Find the length of side *x* :

40°
x
60°
11 m

Round 2

Find the length of side *x* :

13 cm
x
27°
15 cm

Round 3

Find the size of angle *θ* :

9.8 m   *θ*   10.1 m
14.2 m

Round 4

Find the size of angle *θ* :

5.3 m   4.1 m
38° *θ*

Round 5

Find the size of angle *θ* :

8 cm
θ
52°
6 cm

Round 6

Find the length of side *x* :

7.1 m
18°   62°
x

Round 7

Find the length of side *x* :

x   32°
8.6 cm
40°
7 cm

Round 8

Find the size of angle *θ* :

72 m
θ
68 m
24 m

Round 9

A triangle has an angle of 72°, and the side opposite this angle is 12 cm. Another side measures 9 cm. Find the size of the angle opposite the 9 cm side.

Round 10

Ann stands at one point. Ben and Chris are both standing 10 metres from Ann. The angle between the lines from Ben to Ann and Chris to Ann is 40°. How far apart are Ben and Chris standing?

Race 16

Solving Trigonometric Equations

Topics
solving simple trigonometric equations

TEACHER'S NOTES

This is a topic where student discussion with each other can really lead to a greater understanding of the topic. Students can get the principal value from their calculators. It is a good idea to encourage students to sketch the graph of the function they are dealing with, and to use the symmetry of the graphs to find other solutions. The first six rounds all require solutions in the 'usual' range $0° \leq \theta \leq 360°$, and different ranges are used in the last four rounds. All of the equations themselves are quite simple, which is appropriate for GCSE Mathematics, but if your students need extending beyond this, you could replace some of the last rounds with more complex equations.

ANSWERS

Round 1	Round 2
$\theta = \{30°, 150°\}$	$\theta = \{11.54°, 168.46°\}$
Round 3	Round 4
$\theta = \{45.57°, 314.43°\}$	$\theta = \{38.66°, 218.66°\}$
Round 5	Round 6
$\theta = \{210°, 330°\}$	$\theta = \{113.58°, 246.42°\}$
Round 7	Round 8
$\theta = \{30°, 150°\}$	$\theta = \{-243.43°, -63.43°, 116.57°, 296.57°\}$
Round 9	Round 10
$\theta = \{107.46°\}$	$\theta = \{-112.62°, 67.38°\}$

All answers are given to 2 decimal places. Agree with students in advance to what degree of accuracy they should specify answers.

	Round 1		Round 2

Give the solutions to this equation in the range $0° \leq \theta \leq 360°$:
$$\sin\theta = 0.5$$

Give the solutions to this equation in the range $0° \leq \theta \leq 360°$:
$$\sin\theta = 0.2$$

	Round 3		Round 4

Give the solutions to this equation in the range $0° \leq \theta \leq 360°$:
$$\cos\theta = 0.7$$

Give the solutions to this equation in the range $0° \leq \theta \leq 360°$:
$$\tan\theta = 0.8$$

	Round 5		Round 6

Give the solutions to this equation in the range $0° \leq \theta \leq 360°$:
$$\sin\theta = -0.5$$

Give the solutions to this equation in the range $0° \leq \theta \leq 360°$:
$$\cos\theta = -0.4$$

	Round 7		Round 8

Give the solutions to this equation in the range $-180° \leq \theta \leq 180°$:
$$\sin\theta = 0.5$$

Give the solutions to this equation in the range $-360° \leq \theta \leq 360°$:
$$\tan\theta = -2$$

	Round 9		Round 10

Give the solutions to this equation in the range $0° \leq \theta \leq 180°$:
$$\cos\theta = -0.3$$

Give the solutions to this equation in the range $-180° \leq \theta \leq 180°$:
$$\tan\theta = 2.4$$

Race 17

Probability

Topics
simple probability
expectation
sample space diagrams
tree diagrams for independent events
tree diagrams for dependent events

TEACHER'S NOTES

This activity covers a range of probability skills, from basic probability and expectation, to sample space diagrams and tree diagrams. The last two questions involve situations with dependent outcomes. It is also a great way of enforcing the correct format that answers should take, as you can make sure that only probabilities given as fractions or decimals are accepted.

ANSWERS

Round 1	Round 2
0.2	18
Round 3	**Round 4**
$\dfrac{5}{18}$	$\dfrac{1}{12}$
Round 5	**Round 6**
$\dfrac{9}{36} = \dfrac{1}{4}$	$\dfrac{16}{36} = \dfrac{4}{9}$
Round 7	**Round 8**
$\dfrac{80}{144} = \dfrac{5}{9}$	$\dfrac{10}{36} = \dfrac{5}{18}$
Round 9	**Round 10**
$\dfrac{42}{90} = \dfrac{7}{15}$	0.73

Round 1

The probability that Maths United will win their match is 0.6. The probability they lose is the same as the probability they draw. What is the probability they will lose?

Round 2

If the probability that Maths United wins a match is 0.6, how many times would you expect them to win out of 30 matches?

Round 3

Two fair dice are rolled and their scores added together. What is the probability that the score is a number greater than 8?
(Hint: a sample space diagram may help)

Round 4

A fair coin is thrown and a fair die is rolled. What is the probability that the die shows a six and the coin lands tails up?
(Hint: a sample space diagram may help)

Round 5

Millie rolls a die twice and adds the numbers on each die to generate a score. What is the probability that the score is even on both throws?
(Hint: a sample space diagram may help)

Round 6

Marcus rolls a die twice and finds the difference between the numbers on each die to generate a score. What is the probability that the score is a prime number?
(Hint: a sample space diagram may help)

Round 7

Mike has a bag with 4 red counters and 8 blue counters. He takes two counters at random, replacing them after he has noted their colour. What is the probability both counters are the same colour?
(Hint: a tree diagram may help)

Round 8

Molly rolls a die twice, noting whether or not she has rolled a six. What is the probability she rolls a six on one of the rolls, but not both?
(Hint: a tree diagram may help)

Round 9

Matthew is playing 10 songs on shuffle. Each one is selected randomly and won't be selected again once played. He likes 7 of the tracks, and dislikes the other 3. If he listens to 2 tracks, what is the probability he likes one and dislikes the other?
(Hint: a tree diagram may help)

Round 10

The probability that Maria gets up on time is 0.9. If she gets up on time, there is a probability of 0.8 she will walk to work, otherwise she will take the car. If she oversleeps, the probability she takes the car is 0.9. What is the probability she walks?
(Hint: a tree diagram may help)

	Round 1		Round 2
	Round 3		Round 4
	Round 5		Round 6
	Round 7		Round 8
	Round 9		Round 10

	Round 1		Round 2
	Round 3		Round 4
	Round 5		Round 6
	Round 7		Round 8
	Round 9		Round 10